Copyright © 2020 by Discovery Education, Inc. All rights reserved. No part of this work may be reproduced, distributed, or transmitted in any form or by any means, or stored in a retrieval or database system, without the prior written permission of Discovery Education, Inc.

NGSS is a registered trademark of Achieve. Neither Achieve nor the lead states and partners that developed the Next Generation Science Standards were involved in the production of this product, and do not endorse it.

To obtain permission(s) or for inquiries, submit a request to:
Discovery Education, Inc.
4350 Congress Street, Suite 700
Charlotte, NC  28209
800-323-9084
Education_Info@DiscoveryEd.com

ISBN 13: 978-1-68220-788-8

Printed in the United States of America.

5 6 7 8 9 10 CWM 26 25 24 23          B

© Discovery Education | www.discoveryeducation.com

**Acknowledgments**

Acknowledgment is given to photographers, artists, and agents for permission to feature their copyrighted material.

Cover and inside cover art: David Merron / 500px Prime / Getty Images

# Table of Contents

## Unit 2

**Letter to the Parent/Guardian** .................... v

**Animal Sounds** .................................... viii

   Get Started: Wolf Howls ........................ 2

**Unit Project Preview: Making Noise** ............... 4

## Concept 2.1

**Response to Sound** ................................ 6

   **Wonder** ......................................... 8

      Let's Investigate a Baby's Cry ............... 10

   **Learn** .......................................... 16

   **Share** .......................................... 40

## Concept 2.2

**Animal Noises** .................................... 48

   **Wonder** ......................................... 50

      Let's Investigate a Dog Alarm ................ 52

   **Learn** .......................................... 62

   **Share** .......................................... 94

## Unit Project

**Unit Project: Making Noise** ................................. 104

## Grade 1 Resources

Bubble Map ................................................. R1

Safety in the Science Classroom ............................ R3

Vocabulary Flash Cards ..................................... R7

Glossary ................................................... R13

Index ...................................................... R25

# Dear Parent/Guardian,

This year, your student will be using Science Techbook™, a comprehensive science program developed by the educators and designers at Discovery Education and written to the Next Generation Science Standards (NGSS). The NGSS expect students to act and think like scientists and engineers, to ask questions about the world around them, and to solve real-world problems through the application of critical thinking across the domains of science (Life Science, Earth and Space Science, Physical Science).

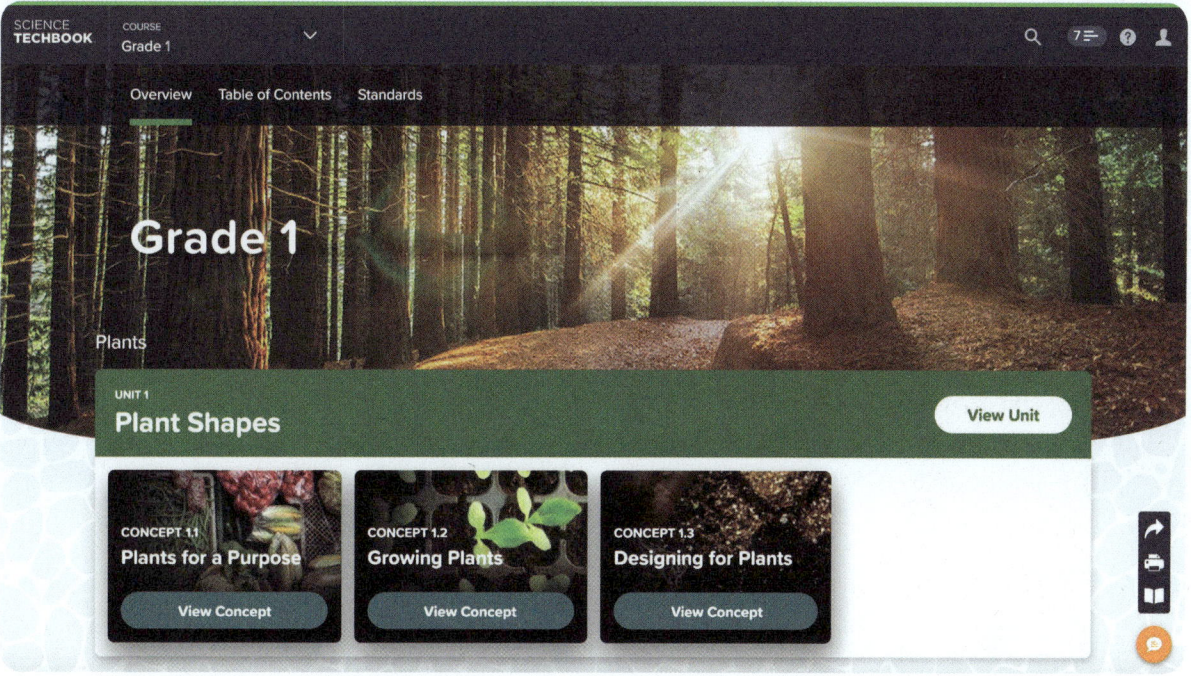

Unit 2: Animal Sounds

Science Techbook is an innovative program that helps your student master key scientific concepts. Students engage with interactive science materials to analyze and interpret data, think critically, solve problems, and make connections across science disciplines. Science Techbook includes dynamic content, videos, digital tools, Hands-On Activities and labs, and gamelike activities that inspire and motivate scientific learning and curiosity.

You and your child can access the resource by signing in to www.discoveryeducation.com. You can view your child's progress in the course by selecting Assignments.

Science Techbook is divided into units, and each unit is divided into concepts. Each concept has three sections: Wonder, Learn, and Share.

**Units and Concepts** Students begin to consider the connections across fields of science to understand, analyze, and describe real-world phenomena.

**Wonder** Students activate their prior knowledge of a concept's essential ideas and begin making connections to a real-world phenomenon and the **Can You Explain?** question.

**Learn** Students dive deeper into how real-world science phenomenon works through critical reading of the Core Interactive Text. Students also build their learning through Hands-On Activities and interactives focused on the learning goals.

**Share** Students share their learning with their teacher and classmates using evidence they have gathered and analyzed during Learn. Students connect their learning with STEM careers and problem-solving skills.

Within this Student Edition, you'll find QR codes and quick codes that take you and your student to a corresponding section of Science Techbook online. To use the QR codes, you'll need to download a free QR reader. Readers are available for phones, tablets, laptops, desktops, and other devices. Most use the device's camera, but there are some that scan documents that are on your screen.

For resources in Science Techbook, you'll need to sign in with your student's username and password the first time you access a QR code. After that, you won't need to sign in again, unless you log out or remain inactive for too long.

We encourage you to support your student in using the print and online interactive materials in Science Techbook on any device. Together, may you and your student enjoy a fantastic year of science!

Sincerely,

**The Discovery Education Science Team**

Unit 2: Animal Sounds | vii

# Unit 2
# Animal Sounds

# Get Started

## Wolf Howls

Pictures in books show wolves howling at the moon. Do wolves howl only at the moon? Do they make noises for other reasons? Do all howls sound the same? At the end of the unit, you will be able to use the information you learned to create your own noise to get the attention of your friend.

**Watch** a video about how wolves use sound to communicate.

Quick Code: us1255s

Wolf Howls

# Think About It

**Look** at the picture. **Think** about the following questions:

- How are parents and their children similar and different?

- How do animal parents and children interact to meet their needs?

- How do animals communicate and make sound?

Penguin and Chick

Unit 2: Animal Sounds | 3

# Unit Project Preview

## Design Solutions Like a Scientist

Quick Code: us1256s

### Hands-On Engineering: Making Noise

In this activity, you will design a solution to better communicate across the school playground.

School Playground

- **SEP** Obtaining, Evaluating, and Communicating Information
- **SEP** Constructing Explanations and Designing Solutions
- **CCC** Structure and Function
- **CCC** Cause and Effect

## Ask Questions About the Problem

You are going to create a device to communicate across the school playground. **Write** some questions you can ask to learn more about the problem. As you work on activities throughout this unit, **write** down answers to your questions.

Unit 2: Animal Sounds | 5

**CONCEPT 2.1**

# Response to Sound

## Student Objectives

By the end of this lesson:

☐ I can find patterns in how young and adult animals look.

☐ I can observe adult animal behaviors that help their young survive.

## Key Vocabulary

☐ animal
☐ communicate
☐ feature
☐ inherit
☐ sound

Quick Code: us1258s

**Activity 1**

# Can You Explain?

How does sound help animals survive?

Quick Code:
us1259s

# 2.1 | Wonder — How does sound help animals survive?

**Activity 2**
# Ask Questions Like a Scientist

Quick Code: us1260s

## A Baby's Cry

**Watch** the video. Then, **answer** the questions.

Let's Investigate A Baby's Cry

### Talk Together

What does it mean when a baby cries?

What do adults do when a baby cries?

What other questions do you have about how a baby **communicates** by crying?

### Your Questions

**SEP** Obtaining, Evaluating, and Communicating Information

2.1 Response to Sound

# 2.1 | Wonder
How does sound help animals survive?

### Activity 3
# Evaluate Like a Scientist

Quick Code: us1261s

## What Do You Already Know About Response to Sound?

## Why Is Sound Important?

Why is each **sound** important? **Draw** a line to show what each sound might tell us.

**Sound**

**Why Is It Important?**

Pay attention!

Time to get up!

Emergency!

Something is done cooking!

## Discussing Basic Needs

What do you need to survive? **Write** or **draw** three things you need.

 **Talk Together**

Now, talk together about basic needs. Talk about the different things you need to survive.

CCC Patterns

2.1 Response to Sound | 13

## 2.1 | Wonder — How does sound help animals survive?

## Which Baby Looks Like Its Parent?

Living things **inherit** traits from their parents. Some baby **animals** look like their parents when they are born. Others do not. However, most animals look like their parents by the time they become adults.

**Look** at the pictures. **Match** each baby animal to its parent.

## Baby Animals

## Adult Animals

2.1 Response to Sound

## 2.1 | Learn   How does sound help animals survive?

### How Are Baby Animals and Adult Animals Different and the Same?

 **Activity 4**
**Observe Like a Scientist**

Quick Code: us1262s

### Lion and Lion Cubs

**Look** at the picture. Then, **answer** the question.

Lion and Lion Cubs

In the chart, **tell** how lion cubs and adult lions are the same and how they are different.

| Same | Different |
|------|-----------|
|      |           |

2.1 Response to Sound

## 2.1 | Learn  How does sound help animals survive?

### Activity 5
### Evaluate Like a Scientist

Quick Code: us1263s

## Who Is My Baby?

**Look** at the pictures of baby animals and their parents. **Match** each parent with its baby.

**Parent Animals**

**Baby Animals**

**SEP** Constructing Explanations and Designing Solutions

**CCC** Patterns

18

**Activity 6**
# Think Like a Scientist

Quick Code: us1264s

## Baby and Adult Animals

In this activity, you will look at photos of baby animals and their parents. You will make observations and take measurements to see how different body parts stay the same or change when animals grow up.

### What materials do you need?
(per group)

- Index cards
- Images of animals (parent)
- Images of animals (offspring)
- Metric ruler
- Pencils

**SEP** Obtaining, Evaluating, and Communicating Information
**CCC** Patterns

2.1 Response to Sound | 19

## 2.1 | Learn   How does sound help animals survive?

## What Will You Do?

**Look** at the pictures. **Circle** the baby in each picture.

**Write** how you know which animal was the baby.

| Animal | Description |
|---|---|
| Giraffe | |
| Sheep | |
| Elephant | |

2.1 Response to Sound

## 2.1 | Learn    How does sound help animals survive?

Now **measure** the body parts with the ruler.

**Write** what you observe in the chart.

| Name of Animal | Name of Body Part | Size of Body Part |
|---|---|---|
| Adult Giraffe | Neck | |
| Baby Giraffe | Neck | |
| Adult Elephant | Ear | |
| Baby Elephant | Ear | |
| Adult Sheep | Leg | |
| Baby Sheep | Leg | |

| | Color of Body Part | Which shape is most similar to the body part: triangle, rectangle, or circle? |
|---|---|---|
| | | |
| | | |
| | | |
| | | |
| | | |
| | | |

2.1 Response to Sound | 23

## 2.1 | Learn — How does sound help animals survive?

**Think About the Activity**

How are baby animals similar to their parents?

How are they different?

**Describe** some ways that living things change as they grow.

_____

_____

_____

_____

**Describe** some ways that living things stay the same as they grow.

_____

_____

_____

_____

## 2.1 | Learn  How does sound help animals survive?

### Why Do Animals Make Sounds?

**Activity 7**
## Observe Like a Scientist

Quick Code: us1265s

### Babies and Adults

**Look** at the pictures of baby and adult animals. Are their **features** the same, similar, or different? **Label** each picture.

Giraffe and Calf

**SEP** Obtaining, Evaluating, and Communicating Information

**CCC** Structure and Function

Elephant with Calf

Sheep and Lamb

2.1 Response to Sound

## 2.1 | Learn  How does sound help animals survive?

 **Activity 8**
**Observe Like a Scientist**

Quick Code: us1266s

## Chickadee Cam

**Watch** the video. Then, **complete** the activities.

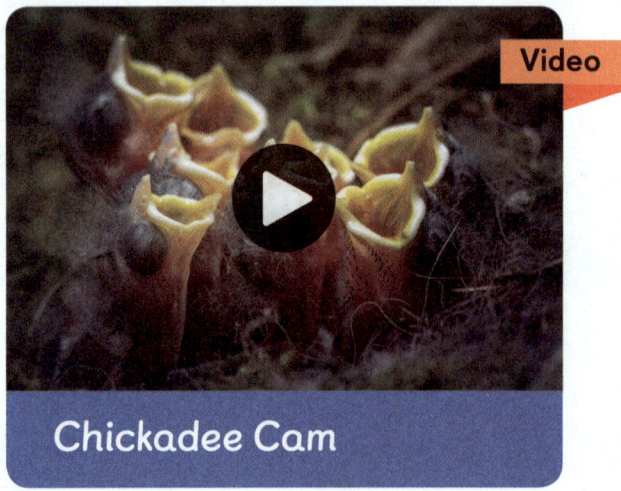
Chickadee Cam

 **Talk Together**

Did you use your voice today to talk to an adult? How is your reason for talking similar to the chickadees' reason for making noise?

**CCC** Cause and Effect

**Draw** what you think the baby chickadees need and what the adult chickadee will do in response to the sounds.

2.1 Response to Sound | 29

**Activity 9**
**Analyze Like a Scientist**

Quick Code: us1267s

## Communicate with Sound

**Write** about a need you might have in the Cause column. Then, **write** the sound you might use to communicate this need in the Effect column.

| Communicate with Sound ||
|---|---|
| Cause | Effect |
|  |  |

How do animals communicate the same needs?

_____

**CCC** Cause and Effect

**Read** the first part of the article on this page. **Underline** the words "make sounds." Then **circle** the reason each sound was made.

### Read Together

# Communicate with Sound

Think about how you used your voice today. Did you need something? Why do you think these animals are making sound?

Animals make sounds for many reasons. Animals make sounds to protect one another. Animals make sounds to warn other animals of other dangerous animals nearby.

Young animals make sounds to let their parent know they need food. What sound do you make when you want something to eat?

**Read Together**

Wolf Howling

Wolves do not howl at the moon, as you may see in pictures. Wolves make howling sounds to protect themselves. Wolves also howl to bring other wolves of their pack to their spot.

**Look** at the picture. **Think** about what you read. Why do wolves howl?

**Activity 10**
# Observe Like a Scientist

## Communication and Social Behavior

**Watch** the video. Then, **answer** the question.

Quick Code: us1268s

Communication and Social Behavior

### 💬 Talk Together

Now talk together about what you learned. Why do animals make sounds?

**SEP** Obtaining, Evaluating, and Communicating Information

2.1 Response to Sound

## How Do Animals Respond to Sounds?

**Activity 11**
**Analyze Like a Scientist**

Quick Code: us1269s

### Responding to Sound

**Read** about sound. Then, **complete** the activity.

Read Together

# Responding to Sound

Children Laugh

Animals make sounds for different reasons. What do other animals do when they hear a sound? What do you do when you hear your friends call your name at recess? You move closer to your friends when they call you over at recess.

**SEP** Obtaining, Evaluating, and Communicating Information
**CCC** Cause and Effect

When baby animals make sounds, adult animals help the baby. A baby bird cheeps. A kitten meows. Adult animals give comfort and basic needs, like food, to the young animals. A kitten meows, curls up, and shakes. The adult cat pulls the kitten in close to make it feel safe.

Lions Growl Softly

In the Cause column, **write** or **draw** when you made a sound based on a need.

In the Effect column, **write** or **draw** how an adult responded.

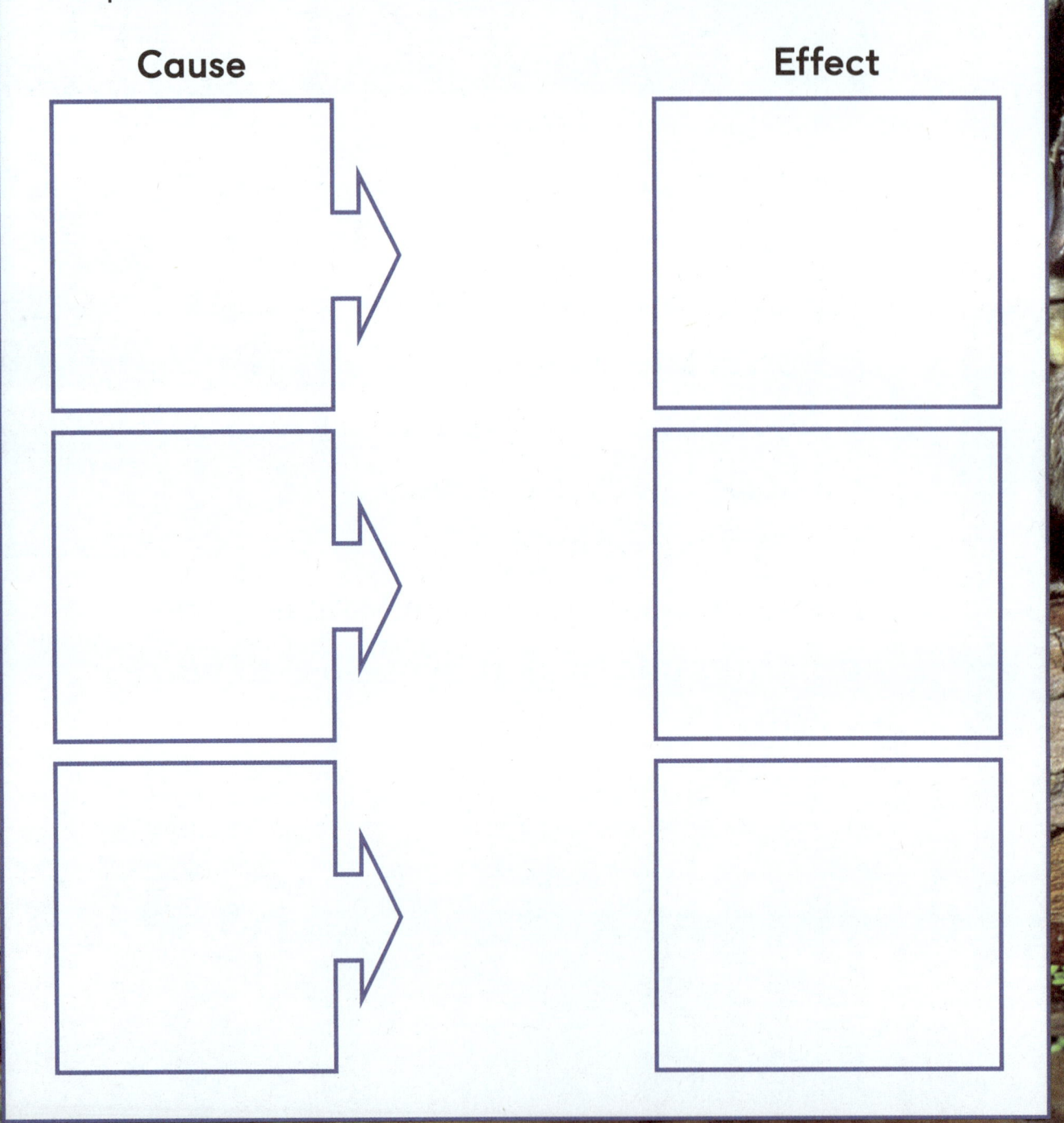

| Cause | Effect |
|---|---|
|  |  |
|  |  |
|  |  |

**Activity 12**
**Observe Like a Scientist**

Quick Code: us1270s

## Penguins

**Watch** the video. **Look** and **listen** for examples of penguins making sounds and how other penguins respond.

[Video: Penguins]

### Talk Together

How do sounds help young penguins and adult penguins find each other?

**CCC** Cause and Effect

2.1 Response to Sound

## 2.1 | Learn — How does sound help animals survive?

### Activity 13
# Evaluate Like a Scientist

Quick Code: us1271s

## What Are the Lions Saying?

**Watch** the video. **Listen** for the sounds the parent lion makes to help the cubs.

Video

Lions and Cubs

**CCC** Cause and Effect

**Match** the images and descriptions of what the lions are saying to the choices. What are the lions trying to say to one another?

Playful, having fun

Warning, go away

Come here

Mother lion growls loudly.

Mother lion growls softly.

Baby lion roars.

Baby and adult lions yip and bark.

2.1 Response to Sound

## 2.1 | Share  How does sound help animals survive?

**Activity 14**

**Record Evidence Like a Scientist**

Quick Code: us1272s

### A Baby's Cry

Now that you have learned about how sound helps animals survive, look again at the video A Baby's Cry. You first saw this in Wonder.

Let's Investigate A Baby's Cry

### Talk Together

How can you describe a baby's cry now? How is your explanation different from before?

**SEP** Constructing Explanations and Designing Solutions

40

Look at the Can You Explain? question. You first read this question at the beginning of the lesson.

> 💬 **Can You Explain?**
> How does sound help animals survive?

Now, you will use your new ideas about babies' cries to answer a question.

1. **Choose** a question. You can use the Can You Explain? question, or one of your own. You can also use one of the questions that you wrote at the beginning of the lesson.

   **Your Question**

2. Then, **use** the sentence starters on the next page to answer the question.

2.1 Response to Sound | 41

## 2.1 | Share — How does sound help animals survive?

Animals use _____

to meet their needs.

I know this because baby animals make sounds to _____

When baby animals make sounds, the adult animal will _____

# STEM in Action

Quick Code: us1273s

**Activity 15**
## Analyze Like a Scientist

### Sounds Help Animals and People

**Read** about how sounds help animals and people. Then, **complete** the activity.

> Read Together

# Sounds Help Animals and People

Did you know that some animals use sound to help them get around? Dolphins are one type of animal that make sounds. Their sounds bounce off of their surroundings. Then the sounds come back to the dolphin. The dolphins use the sounds to help them move around.

Dolphin

2.1 Response to Sound | 43

**Read Together**

The dolphin uses sound to find food. People use sound to help them find objects, too. Captains of ships use machines to send out sounds. The sounds bounce off objects. Then, the sounds can help the pilots find obstacles like icebergs. The sounds help them travel safely.

Dolphins

Some scientists study icebergs. Some of this iceberg is above the water. Most of the iceberg is below the water. Scientists use sound to find out how deep the iceberg is. They can use sound to measure the iceberg.

They measure how far it rises above the water. They measure how deep it is below the water. Then they can work out how tall the iceberg is.

Ship and Iceberg

2.1 Response to Sound | 45

How tall do you think this iceberg is?
**Circle** the height.

Iceberg

75 cm    75 m    75 km

SEP  Obtaining, Evaluating, and Communicating Information

**Activity 16**
# Evaluate Like a Scientist

Quick Code: us1274s

## Review: Response to Sound

**Think** about what you have read and seen. What did you learn?

**Draw** what you have learned.

Then, **tell** someone else about what you learned.

### 💬 Talk Together

Think about what you saw in Get Started. Use your new ideas to discuss how sound helps animals survive.

CONCEPT
2.2

# Animal Noises

## Student Objectives

By the end of this lesson:

- ☐ I can design, build, play, and compare devices that use sound to communicate over a distance.

- ☐ I can use an investigation to explain how sounds are created by different materials.

- ☐ I can use an investigation to gather evidence on how sounds affect materials.

## Key Vocabulary

- ☐ air
- ☐ sense
- ☐ vibration
- ☐ volume
- ☐ wave

Quick Code: us1276s

2.2 Animal Noises

## Activity 1
# Can You Explain?

How do animals use their body parts to make sounds?

Quick Code: us1277s

2.2 Animal Noises | 51

2.2 | **Wonder**  How do animals use their body parts to make sounds?

**Activity 2**
**Ask Questions Like a Scientist**

Quick Code: us1278s

## Dog Alarm

**Watch** the video. **Answer** the questions.

Let's Investigate a Dog Alarm

**SEP** Developing and Using Models

## Talk Together

How do you think a dog can make different sounds?

**Draw** a picture of how a dog makes sound.

What other questions do you have about how a dog makes different sounds?

Your Questions

2.2 Animal Noises | 53

### Activity 3
# Analyze Like a Scientist

Quick Code: us1279s

## Alarm Dog

**Listen** as your teacher reads. **Think** about answers to the questions you hear.

**Read Together**

# Alarm Dog

Have you ever heard an ambulance or police car siren? Is the sound loud or soft?

Fire alarms in schools help you and your classmates know if there is a fire in the building.

Some dogs can make themselves sound like an alarm or siren!

Fire Alarm

**SEP** Obtaining, Evaluating, and Communicating Information

54

Can you make a sound like a siren?

Can you whistle?

**Boy Whistling**

When you whistle, you push **air** out of your mouth.

**Draw** something that makes a loud sound.

## 2.2 | Wonder — How do animals use their body parts to make sounds?

### Activity 4
### Observe Like a Scientist

Quick Code: us1280s

## Sounds All Around

**Listen** to the song. Then, **answer** the questions.

Audio

Sounds All Around

### Talk Together

How many sounds can you hear? How do you think the sounds were made?

**Listen** to the sounds again. **Write** or **draw** a description of the different sounds you hear.

2.2 Animal Noises | 57

## 2.2 | Wonder How do animals use their body parts to make sounds?

### Activity 5
### Observe Like a Scientist

Quick Code: us1281s

**Sorting Sounds**

**Watch** the video. Then, **answer** the questions.

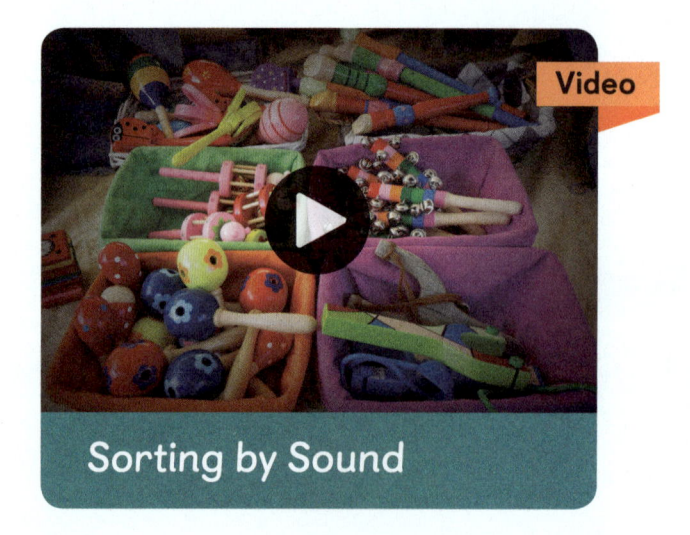

Sorting by Sound

### Talk Together

Now talk with your class. What sounds did you hear? How did the students sort the sounds?

CCC Patterns

**Activity 6**

# Evaluate Like a Scientist

Quick Code: us1282s

## What Do You Already Know About Animal Noises?

### Discussing Sounds

Can you think of different ways to make sounds? What do these ways of making sound have in common?

## 2.2 | Wonder — How do animals use their body parts to make sounds?

### How Do They Vibrate?

How do the different instruments make a sound? **Write** Blow, Hit, or Pull under each picture to show how the instruments make sound.

60

2.2 Animal Noises | 61

## 2.2 | Learn  How do animals use their body parts to make sounds?

### How Are Sounds Made?

**Activity 7**
**Observe Like a Scientist**

Quick Code: us1283s

## Musical Instruments

How do these instruments make sounds?

**Look** at the picture.

Musical Instruments

## Talk Together

Now talk together about what you see. What do the instruments remind you of? What do you think the instruments sound like?

2.2 Animal Noises

## 2.2 | Learn — How do animals use their body parts to make sounds?

**Activity 8**
# Think Like a Scientist

Quick Code: us1284s

## Sound Exploration

In this activity, you will use your **senses** of hearing and touch to explore sound energy.

### What materials do you need?
(per group)

- Instruments such as drums, maracas, recorders
- CD player
- Earmuffs
- Variety of surfaces and objects to make sounds (ridged cardboard, metal can, stick or pencil)
- Cardboard
- Can, metal
- Pencils

**SEP** Constructing Explanations and Designing Solutions

**CCC** Cause and Effect

## What Will You Do?

Visit the Music Exploration Center. Play the instruments and **listen** to the sounds.

Visit the Sound Exploration Center. **Create** sounds with the objects. Use words to describe the sounds.

What senses will you use to explore sound?

_____

Visit the Music Exploration Center. **Draw** pictures of the instruments you saw. What words **describe** the sounds you made?

## 2.2 | Learn   How do animals use their body parts to make sounds?

Visit the Sound Exploration Center. Use the different materials to make sounds. **Draw** pictures of the objects you use to make sound. What words **describe** the sounds?

# Think About the Activity

What two body parts did you use to explore sound?

How did those body parts help you explore sound?

2.2 Animal Noises | 67

## 2.2 | Learn — How do animals use their body parts to make sounds?

### Activity 9
### Evaluate Like a Scientist

Quick Code: us1285s

## Describing Sounds

**Draw** a line from each picture to the word that tells about the sound it makes.

- ring
- buzz 
- rumble 
- squeal 

**CCC** Structure and Function

**Activity 10**

# Design Solutions Like a Scientist

Quick Code: us1286s

## Hands-On Engineering: Vibrating Instruments

In this activity, you will work with a small group to design your own instrument using everyday items. Use what you know about sound to make your instrument even better!

## Ask Questions About the Problem

You are going to design an instrument. What questions do you need to ask before you design? What do you need to know about first?

**SEP** Planning and Carrying Out Investigations

**CCC** Cause and Effect

2.2 Animal Noises

## 2.2 | Learn   How do animals use their body parts to make sounds?

### What materials do you need?
(per group)

**For Part 1:**
- Objects such as a metal can, seeds, rubber bands, a small box, plastic straws, a comb, balloons, an aluminum foil pan, a pin, a paper cup, a book, clear tape, scissors, glue, a plastic bottle, a plastic spoon, and a pitcher of water

**For Part 2:**
- Objects such as cotton balls, a whoopee cushion, a whistle, and a metric ruler

### What Will You Do?

With your group, **look** at the materials from your teacher. **Think** about what you have learned about sound. **Draw** one or two ideas for your instrument.

**Build** your instrument. What is the name of your instrument?

**Draw** your instrument.

2.2 Animal Noises

## 2.2 | Learn — How do animals use their body parts to make sounds?

### Think About the Activity

Why do different instruments make different sounds?

---

### 💬 Talk Together

Now, talk about how you could design an investigation to show that vibrating materials make sound.

Do the investigation you talked about. **Draw** your observations from your investigation.

Vibrating materials make sound. Do your observations support this claim?

### Activity 11
# Analyze Like a Scientist

Quick Code: us1287s

## Animal Noises

**Read** the article. Then, **complete** the activity.

🔖 Read Together

# Animal Noises

Many animals make noise. How do different animals make noise? What body parts do they use?

Cricket

A cricket is an insect with wings. It can rub its wings together. The rubbing makes a *chirp* sound. Male crickets make this noise for female crickets.

**CCC** Structure and Function

74

**Rattlesnake**

Rattlesnakes have special tails. When they move their tails, the sections vibrate and make a buzzing sound. Every year, a rattlesnake's tail gets bigger. Old rattlesnakes have big tails!

**Frog**

Frogs make a *ribbit* noise. To make this sound, frogs have air in their throats. The air is stored in an air sac. When air goes in and out of this sac, it makes noise. If you look closely, you can see the air sac.

2.2 Animal Noises

**Read Together**

Elephants

Elephants make sound with their trunks. They can push air through their trunks. The air vibrates and can make different sounds. It is almost like playing a trumpet!

**Choose** two animals you read about. **Make** a chart to compare how the animals make noise.

| Animal | Noise |
|--------|-------|
|        |       |

2.2 Animal Noises

## 2.2 | Learn  How do animals use their body parts to make sounds?

**Activity 12**
## Investigate Like a Scientist

Quick Code: us1288s

### Hands-On Investigation: Making Vibrations

In this activity, you will make sounds with materials. You will observe to see how materials vibrate. You will observe how sound affects materials.

### Make a Prediction

How can you use sound to make an object move?

**SEP** Planning and Carrying Out Investigations
**CCC** Cause and Effect

## What materials do you need?
(per group)

- Chart paper
- Tuning fork
- Paper
- Drum
- Paper clips
- Plastic container, 12 oz
- Water

## What Will You Do?

With your group, think about your plan and the materials you have. **Draw** or **write** how you will make sounds for each material: tuning fork, drum, water.

2.2 Animal Noises

## 2.2 | Learn — How do animals use their body parts to make sounds?

### Think About the Activity

What happened when you struck the tuning fork and drum?

| Cause | Effect |
|---|---|
| I struck the tuning fork. | |
| I placed paper near the tuning fork. | |
| I hit the drum. | |
| I put paper clips on the drum. | |

How do materials make sound? Tell how you know.

How does sound affect a material? Tell how you know.

## 2.2 Learn — How do animals use their body parts to make sounds?

### How Does an Animal Hear Sounds?

**Activity 13**
**Observe Like a Scientist**

Quick Code: us1289s

### Animals Making and Hearing Sounds

**Look** at the picture of the elephant. Then, **answer** the question.

Elephant

### Talk Together

What body part helps an elephant hear sounds?

**SEP** Constructing Explanation and Designing Solutions
**CCC** Structure and Function

**Activity 14**

**Analyze Like a Scientist**

Quick Code: us1290s

## Hearing Humans

**Read** the article. **Circle** the body parts you use to hear.

Read Together

# Hearing Humans

Animals use their body parts to make sound. What body part do animals use to hear sound? What body part do you use to hear sound?

You use your ears to hear sound. How does that work? You can hear sounds that are close and far away.

If sound can make objects vibrate, how do you think your ear works? What parts of your ear move? One part of the ear is the eardrum. Why do they call it an eardrum?

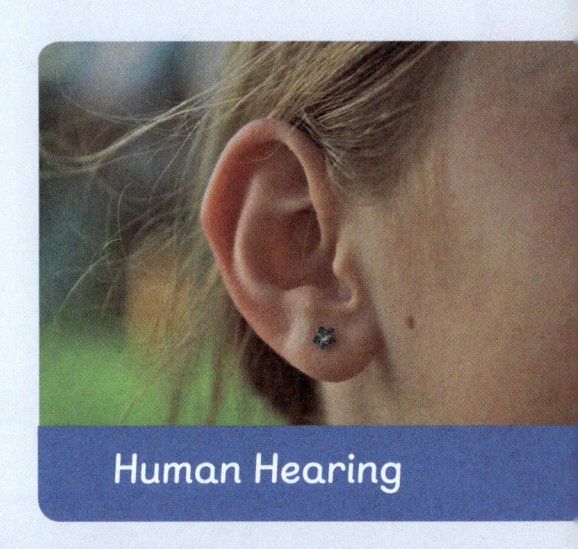

Human Hearing

2.2 Animal Noises | 83

**Explain** how an animal's ear works to hear sound.
**Draw** or **write** your answer.

SEP  Constructing Explanations and Designing Solutions

Activity 15

## Observe Like a Scientist

Quick Code: us1291s

## Inside Your Ear

**Watch** the video. **Look** at the image.

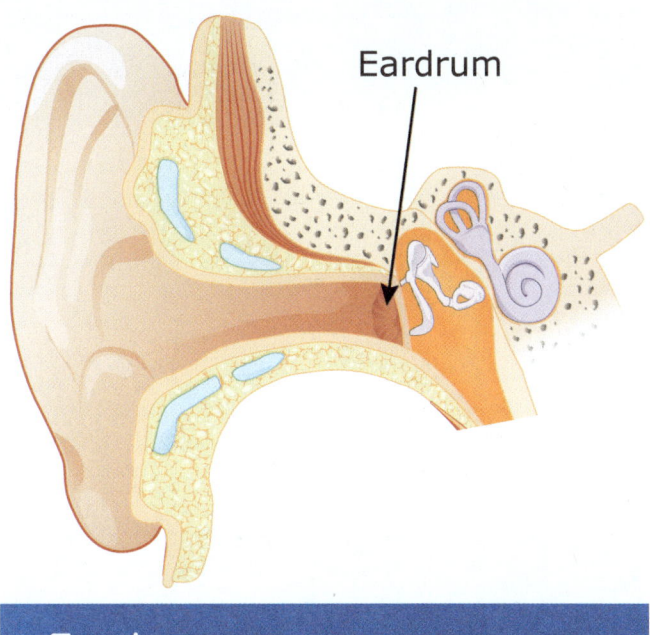

Video

Inside Your Ear

Eardrum

Eardrum

2.2 Animal Noises | 85

## 2.2 | Learn — How do animals use their body parts to make sounds?

### 💬 Talk Together

Now, talk together about how an ear hears sound.

**Look** at the drawing you made in Activity 14. What changes would you like to make to the drawing? **Make** a new drawing.

**SEP** Constructing Explanations and Designing Solutions

**Activity 16**

# Evaluate Like a Scientist

Quick Code: us1292s

## Sound Energy

**Name** an example of sound energy that you hear at school. How does this sound travel to your ear? **Draw** and **label** a picture to show the sound traveling to your ear.

**SEP** Constructing Explanations and Designing Solutions

2.2 Animal Noises | 87

## 2.2 | Learn — How do animals use their body parts to make sounds?

**Activity 17**
**Investigate Like a Scientist**

Quick Code: us1293s

## Hands-On Investigation: Making a Telephone

In this activity you will use a model telephone to explore how sound **waves** travel.

## Make a Prediction

You are going to make a telephone! How can you build a model telephone? **Write** or **draw** your predictions.

- **SEP** Planning and Carrying Out Investigations
- **SEP** Asking Questions and Defining Problems
- **CCC** Energy and Matter

## What materials do you need?
(per group)

- Paper cups, 360 mL, 2
- Pencils
- String
- Paper clips, large

*HANDS-ON INVESTIGATION*

**Complete** the sentence frame.

Sound is made when objects

2.2 Animal Noises

## 2.2 | Learn — How do animals use their body parts to make sounds?

**What Will You Do?**

**Make** your telephone using the materials from your teacher.

**Draw** how you made your telephone.

**Use** the telephone to talk to your partner. What happened?

## 2.2 | Learn — How do animals use their body parts to make sounds?

**Think About the Activity**

How does sound travel through your telephone?

**Explain** what problem your model telephone can solve.

**Describe** how your model telephone is like a real telephone.

## 2.2 | Share  How do animals use their body parts to make sounds?

**Activity 18**

**Record Evidence Like a Scientist**

Quick Code: us1294s

### Dog Alarm

Now that you have learned more about how animals make sounds and hear sounds, look again at the Dog Alarm video. You first saw this in Wonder.

Let's Investigate a Dog Alarm

**Talk Together**

How can you describe the Dog Alarm video now? How is your explanation different from before?

**SEP** Constructing Explanations and Designing Solutions

Look at the Can You Explain? question. You first read this question at the beginning of the lesson.

> 💬 **Can You Explain?**
>
> How do animals use their body parts to make sounds?

Now, you will use your new ideas about dog alarms to answer a question.

1. **Choose** a question. You can use the Can You Explain? question, or one of your own. You can also use one of the questions that you wrote at the beginning of the lesson.

**Your Questions**

2. Then, **use** the sentence starters to **answer** the question.

2.2 Animal Noises | 95

## 2.2 | Share  How do animals use their body parts to make sounds?

Sounds are made when

An example of how animals make sounds is

People make sounds by

**Activity 19**

# Analyze Like a Scientist

Quick Code: us1295s

## Making Sound for a Job

Read the article. Then, **complete** the activity.

**Read Together**

# Making Sound for a Job

Can you imagine a job in which you get paid to make sounds? Musicians use instruments or their voices to make music. Music is sound that people like to hear. People buy recordings of music by musicians they like. They buy tickets to go to concerts where the musician is playing.

Drum Set

**SEP** Using Mathematics and Computational Thinking

**SEP** Obtaining, Evaluating, and Communicating Information

98 | Discovery Education

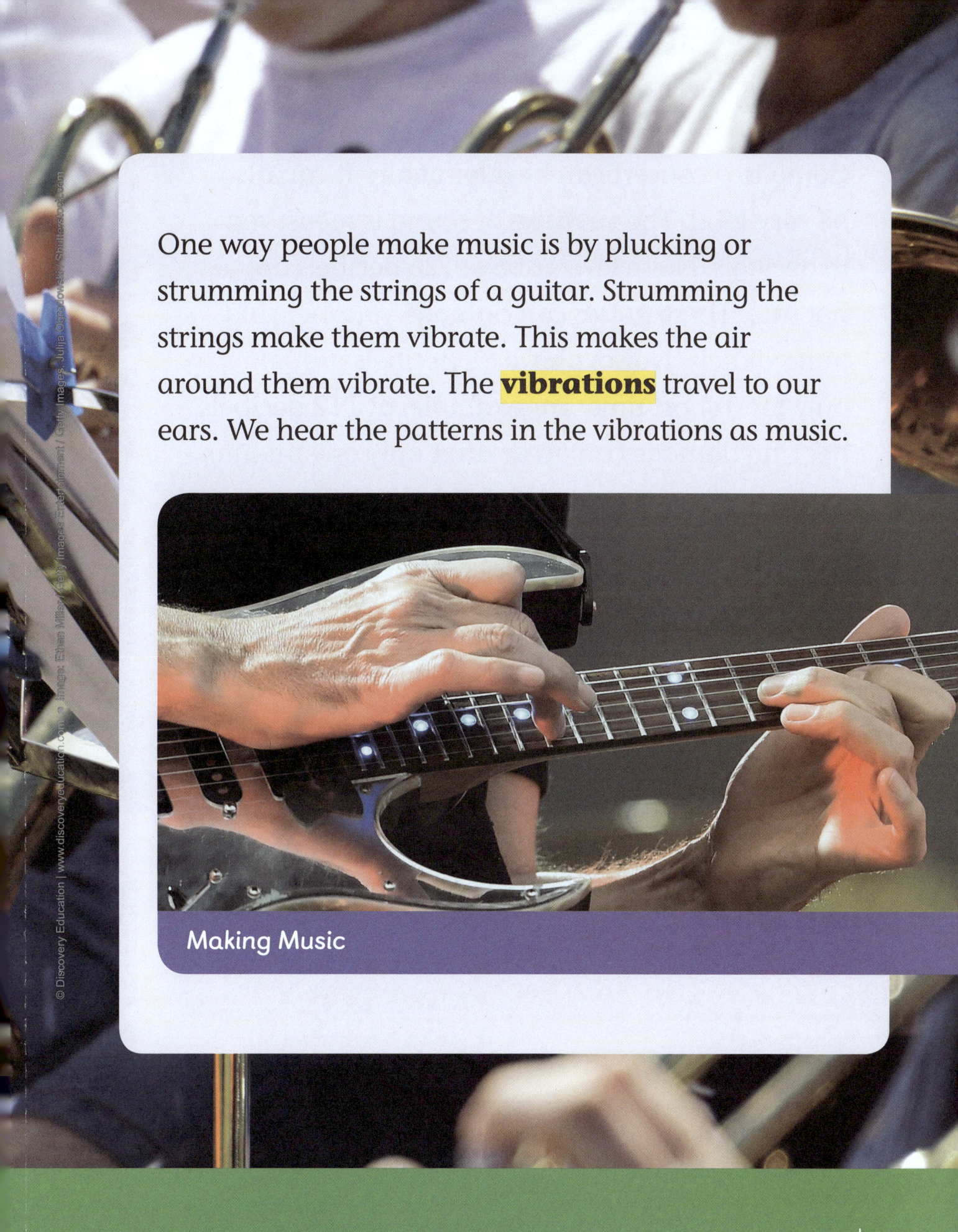

One way people make music is by plucking or strumming the strings of a guitar. Strumming the strings make them vibrate. This makes the air around them vibrate. The **vibrations** travel to our ears. We hear the patterns in the vibrations as music.

Making Music

2.2 Animal Noises

**Read Together**

Going to a concert can be a lot of fun. It can also be very loud. The **volume** of sound is measured in decibels. Noise levels above 120 decibels are not safe. These noises can damage your hearing immediately. Noises above 85 decibels are safe only for a short time. For example, at 112 decibels, hearing damage can occur after only one minute.

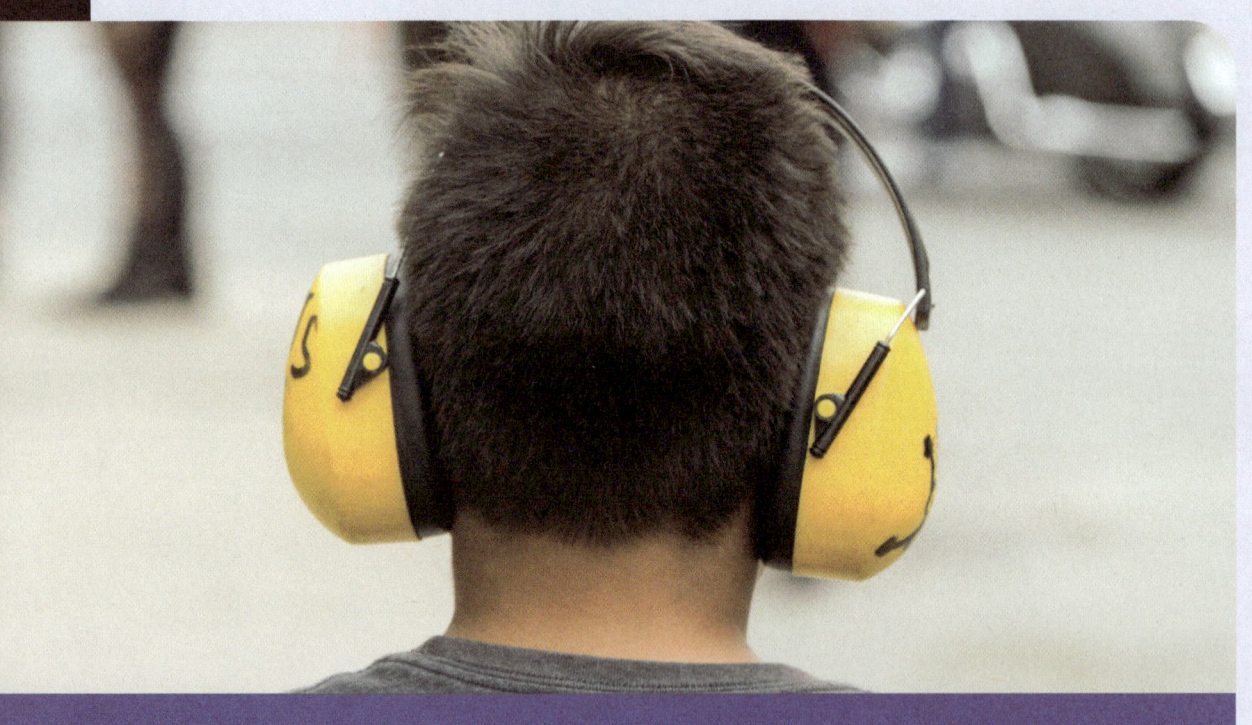

Protection from Loud Noise

**Look** at the decibel table. Use green to **color** the sounds that are safe. Use yellow to **color** the sounds that are safe for a short time. Use red to **color** the sounds that are not safe.

| |
|---|
| 10 normal breathing |
| 50 rainfall |
| 60 normal talking |
| 90 lawn mower |
| 110 car horn |
| 115 police siren |
| 120 jet engine |
| 130 jackhammer |
| 180 rocket launch |

## 2.2 | Share  How do animals use their body parts to make sounds?

**Activity 20**
# Evaluate Like a Scientist

Quick Code: us1296s

## Review: Animal Noises

**Think** about what you have read and seen.

What did you learn?

**Draw** what you have learned.

Then, **tell** someone else about what you learned.

## 💬 Talk Together

Think about what you saw in Get Started. Use your new ideas to discuss how animals make sounds.

**SEP** Obtaining, Evaluating, and Communicating Information

2.2 Animal Noises

# Unit Project

## Design Solutions Like a Scientist

Quick Code: us1297s

### Hands-On Engineering: Making Noise

In this activity, you will design an object that can be used to communicate across the school playground.

School Playground

- **SEP** Constructing Explanations and Designing Solutions
- **SEP** Obtaining, Evaluating, and Communicating Information
- **CCC** Structure and Function
- **CCC** Cause and Effect

## What materials do you need?
(per group)

- Can, metal
- Metal teaspoon, wooden spoon, plastic spoon
- Wood block
- Metal object
- Beakers, 250 mL
- Aluminum foil pan, 22.5 cm
- Wooden board

*HANDS-ON ENGINEERING*

## What Will You Do?

**Draw** a picture of your device design.
How will it work to communicate across a playground?

# Unit Project

What materials will you need for your design?

**Test** your design. **Draw** a picture or use words to show how you tested your design.

# Think About the Activity

**Write** or **draw** your answers to the questions in the chart.

How well did your design help communicate across the school playground?

How could you improve your design?

| What Didn't Work? | What Worked? |
|---|---|
|  |  |

**What Could Work Better?**

**What Is Your Final Design?**

# Grade 1 Resources

- Bubble Map
- Safety in the Science Classroom
- Vocabulary Flash Cards
- Glossary
- Index

Name _____

## Bubble Map

Can You Explain?
Question:

Bubble Map | R1

# Safety in the Science Classroom

Following common safety practices is the first rule of any laboratory or field scientific investigation.

## Dress for Safety

One of the most important steps in a safe investigation is dressing appropriately.

- Splash goggles need to be kept on during the entire investigation.

- Use gloves to protect your hands when handling chemicals or organisms.

- Tie back long hair to prevent it from coming in contact with chemicals or a heat source.

- Wear proper clothing and clothing protection. Roll up long sleeves, and if they are available, wear a lab coat or apron over your clothes. Always wear closed-toe shoes. During field investigations, wear long pants and long sleeves.

Safety Goggles

## Be Prepared for Accidents

Even if you are practicing safe behavior during an investigation, accidents can happen. Learn the emergency equipment location in your classroom and how to use it.

- The eye and face wash station can help if a harmful substance or foreign object gets into your eyes or onto your face.

- Fire blankets and fire extinguishers can be used to smother and put out fires in the laboratory. Talk to your teacher about fire safety in the lab. He or she may not want you to directly handle the fire blanket and fire extinguisher. However, you should still know where these items are in case the teacher asks you to retrieve them.

Most importantly, when an accident occurs, immediately alert your teacher and classmates. Do not try to keep the accident a secret or respond to it by yourself. Your teacher and classmates can help you.

Fire Extinguisher

## Practice Safe Behavior

There are many ways to stay safe during a scientific investigation. You should always use safe and appropriate behavior before, during, and after your investigation.

- Read all of the steps of the procedure before beginning your investigation. Make sure you understand all the steps. Ask your teacher for help if you do not understand any part of the procedure.

- Gather all your materials and keep your workstation neat and organized. Label any chemicals you are using.

- During the investigation, be sure to follow the steps of the procedure exactly. Use only directions and materials that have been approved by your teacher.

- Eating and drinking are not allowed during an investigation. If asked to observe the odor of a substance, do so using the correct procedure known as wafting, in which you cup your hand over the container holding the substance and gently wave enough air toward your face to make sense of the smell.

- When performing investigations, stay focused on the steps of the procedure and your behavior during the investigation. During investigations, there are many materials and equipment that can cause injuries.

- Treat animals and plants with respect during an investigation.

- After the investigation is over, appropriately dispose of any chemicals or other materials that you have used. Ask your teacher if you are unsure of how to dispose of anything.

- Make sure that you have returned any extra materials and pieces of equipment to the correct storage space.

- Leave your workstation clean and neat. Wash your hands thoroughly.

# Vocabulary Flash Cards

## air

an invisible gas that is all around us; living things, such as plants and animals need it to breathe and grow

## animal

a living thing that moves around to look for food, water, or shelter, but can't make its own food

## communicate

to give and get information, messages, or ideas

## energy

the ability to do work or make something change

Vocabulary Flash Cards | R7

## feature

a thing that describes what something looks like; part of something

## inherit

to receive a characteristic from one's parents

## sense

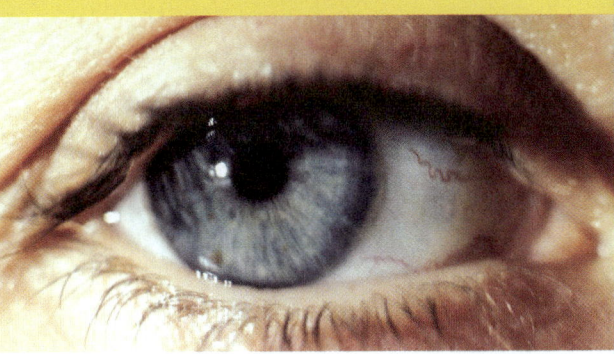

sight, hearing, smell, taste, or touch

## sound

anything that people or animals can hear with their ears

Vocabulary Flash Cards | R9

## trait

a characteristic that you get from one of your parents

## vibration

the rapid movement of an object back and forth

## volume

the loudness of a sound

## wave

the way sound moves through the air

Vocabulary Flash Cards | R11

# Glossary

## English ——— A ——— Español

**absorb**
to take in or soak up

**absorber**
tomar o captar

**air**
an invisible gas that is all around us; living things, such as plants and animals, need it to breathe and grow

**aire**
gas invisible que nos rodea; todos los seres vivos, como las plantas y los animales, lo necesitan para respirar y crecer

**animal**
a living thing that moves around to look for food, water, or shelter, but can't make its own food

**animal**
ser vivo que se mueve para buscar alimento, agua o refugio, pero no puede producir su propio alimento

**axis**
a real or imaginary line through the center of an object; the object turns around it

**eje**
línea real o imaginaria que pasa por el centro de un objeto; el objeto gira alrededor de ella

## B

**binoculars**
a device that is put up to your eyes so you can see far away

**binoculares**
dispositivo que se pone sobre los ojos para poder ver lejos

## C

**collect**
to gather

**recolectar**
reunir

**communicate**
to give and get information, messages, or ideas

**comunicarse**
dar y recibir información, mensajes o ideas

**constellation**
a particular area of the sky; a group of stars

**constelación**
área particular del cielo; grupo de estrellas

## E

**Earth**
the third planet from the Sun; the planet on which we live (related words: earthly; earth - meaning soil or dirt)

**Tierra**
tercer planeta desde el Sol; planeta en el cual vivimos (palabras relacionadas: terrenal; tierra en el sentido de suelo o barro)

**edible**
able to be eaten as a food

**comestible**
que se puede comer como alimento

**energy**
the ability to do work or make something change

**energía**
habilidad de trabajar o producir un cambio

**engineer**
a person who designs something that may be helpful to solve a problem

**ingeniero**
persona que diseña algo que puede ser útil para resolver un problema

## F

**feature**
a thing that describes what something looks like; part of something

**rasgo**
cosa que describe cómo se ve algo; parte de algo

**flower**
the plant part that blooms with colorful petals and beautiful smells and holds the part of the plant that makes the seeds

**flor**
parte de la planta que florece con pétalos de colores y aromas agradables y contiene la parte que produce las semillas

**fruit**
the plant part that contains seeds and grows from a flowering plant

**fruta**
parte de la planta que contiene semillas y crece de una planta en flor

## I

**inherit**
to receive a characteristic from one's parents

**heredar**
recibir una característica de los padres de alguien

## L

**leaf**
the part of the plant that grows off the stem and collects sunlight for the plant to make food

**hoja**
parte de la planta que crece desde el tallo y reúne luz solar para que la planta produzca alimento

**light**
a form of energy that makes it possible for our eyes to see

**luz**
forma de energía que hace posible ver con los ojos

## M

**material**
things that can be used to build or create something

**material**
cosas que se pueden usar para construir o crear algo

**measure**
to find the amount, the weight, or the size of something

**medir**
hallar la cantidad, el peso o el tamaño de algo

**moon**
any object that goes around a planet

**luna**
cualquier objeto que gira alrededor de un planeta

## N

**nutrient**
something in food that helps people, animals, and plants live and grow

**nutriente**
algo en los alimentos que ayuda a las personas, los animales y las plantas a vivir y crecer

## O

**observe**
to watch closely

**observar**
mirar atentamente

**opaque**
when no light gets through something, like wood or metal

**opaco**
cuando no pasa luz a través de algo, como la madera o el metal

**orbit**
to travel in a circular path around something

**orbitar**
viajar un recorrido circular alrededor de algo

## P

**plant**
a living thing made up of cells that needs water and sunlight to survive

**planta**
ser vivo formado por células que necesita agua y luz solar para sobrevivir

**position**
a place where a person or a thing is located

**posición**
lugar donde se encuentra una persona o cosa

**property**
a characteristic of something

**propiedad**
característica de algo

## R

**reflect**
when something like light or heat bounces off a surface

**reflejar**
cuando algo como la luz o el calor rebota en una superficie

**rotate**
to turn around a center point; to spin

**rotar**
girar alrededor de un punto central; dar vueltas

## S

**seed**
the small part of a flowering plant that grows into a new plant

**semilla**
parte pequeña de una planta en flor que se convierte en una nueva planta

**seedling**
a baby plant that starts from a seed

**plántula**
planta joven que crece de una semilla

**sense**
sight, hearing, smell, taste, or touch

**sentido**
visión, audición, olfato, gusto o tacto

**soil**
dirt that covers Earth, in which plants can grow and insects can live

**suelo**
tierra que cubre nuestro planeta en la que pueden crecer plantas y vivir insectos

**sound**
anything that people or animals can hear with their ears

**sonido**
todo lo que las personas o los animales pueden oír con los oídos

**source**
the start or the cause of something

**fuente**
el comienzo o la causa de algo

Glossary | R21

**star**
a burning ball of gas in space

**stem**
the part of the plant that grows up from the roots and holds up the leaves and flowers

**structure**
a part of an organism; the way parts are put together

**sun**
any star around which planets revolve

**system**
a group of parts that go together to make something work

**estrella**
bola ardiente de gas en el espacio

**tallo**
parte de la planta que crece hacia arriba desde las raíces y sostiene las hojas y las flores

**estructura**
parte de un organismo; la forma en que se unen las partes

**sol**
toda estrella alrededor de la cual giran los planetas

**sistema**
grupo de partes que se combinan para hacer que algo funcione

## T

**technology**
inventions that were developed to solve problems and make things easier

**tecnología**
inventos que se desarrollaron para resolver problemas y hacer más fáciles las cosas

**tendril**
a long, thin stem that wraps around things as it grows

**zarcillo**
tallo largo y delgado que se enrosca alrededor de cosas a medida que crece

**trait**
a characteristic that you get from one of your parents

**rasgo**
característica que se recibe de uno de los padres

**translucent**
when some light gets through and what is on the other side might not be very clear, like fog

**translúcido**
cuando pasa algo de luz y lo que hay del otro lado puede no ser muy transparente, como la niebla

**transparent**
when light passes through and you can see clearly, such as clean water and air

**transparente**
cuando pasa la luz y se puede ver con claridad, como el agua limpia y el aire

--- **V** ---

**vibration**
the rapid movement of an object back and forth

**vibración**
rápido movimiento de un objeto adelante y atrás

**volume**
the loudness of a sound

**volumen**
la intensidad de un sonido

--- **W** ---

**water**
a clear liquid that has no taste or smell

**agua**
líquido transparente que no tiene sabor ni olor

**wave**
the way sound moves through the air

**onda**
manera en la que el sonido viaja por el aire

# Index

## A

Adult animals, 14–27, 35, 42
Air
    elephant noises made with, 76
    frog noises made with, 75
    needed to whistle, 55
Alarms
    and dogs, 52–55, 94
    sounds made by, 54
Analyze Like a Scientist, 30, 34–36, 43–46, 54–55, 74–77
Animal parts
    making sound with, 50
    measurements of, 19–25
Animals
    babies and adults, 14–27, 35, 42
    ears of, 84
    noises made by, 59, 74–77, 102–103
    sounds helping, 43–46
    sounds made by, 96

Ask Questions Like a scientist, 10–11, 52–53

## B

Babies, crying of, 10–11, 40–42
Baby animals, 14–27, 35, 42
Body parts
    exploring sounds with, 67
    for hearing, 83–84
    used to make noise, 74

## C

Can You Explain?, 8, 50
Chickadees, 28–29
Communication
    by babies, 11
    across the playground, 4–5, 104–107
    and social behavior, 33
    with sound, 30–32, 35–36, 39, 42
Comparing, how animals make noises, 77
Concerts, 100

Crickets, 74
Crying, of babies, 10–11, 40–42

# D

Decibels, 100–101
Design Solutions Like a Scientist, 4–5, 69–73, 104–107
Dogs
    and alarms, 52–55, 94
    sounds made by, 54
Dolphins, 43–44

# E

Eardrum, 83
Ears
    of animals, 84
    of humans, 83
    inside of, 85–86
Elephants
    babies and adults, 21–23
    hearing of, 82
    sounds make by, 76
Energy of sound, 64–67, 87
Evaluate Like a Scientist, 12–15, 18, 38–39, 47, 59–61, 68, 87, 102–103
Exploring, sound energy, 64–67

# F

Features, of animals, 26
Food, finding, 44
Frogs, 75

# G

Giraffes, 21–23
Guitar, 99

# H

Hearing
    body parts for, 82
    of humans, 83–84
Howling, 2, 32
Humans
    hearing of, 83–84
    sounds helping, 43–46
    sounds made by, 97

## I

Icebergs, 44–46
Inheritance, of traits, 14
Investigate Like a Scientist, 78–81, 88–93

## L

Lions, 16–17, 38–39
Living things, 25

## M

Materials
    making instruments out of, 70
    for making phone, 89, 106
    vibrating, 79–81
Measurements
    of animal parts, 19–25
    of icebergs, 45–46
Movement, 43
Music, 98–100
Musical instruments
    building, 69–73
    sounds created with, 60, 62–63, 65
    used by musicians, 98

Music Exploration Center, 64–65

## N

Needs
    of humans, 13
    making sounds to communicate, 30, 35–36, 42
Noises
    made by animals, 59, 74–77, 102–103
    making, 4–5, 104–107
    safe and unsafe levels of, 100–101

## O

Objects, making sounds with, 65, 66, 79–81
Observations, about animals, 19–25
Observe Like a Scientist, 16–17, 26–27, 28–29, 33, 37, 56–57, 58, 62–63, 82, 85–86

# P

Parents, responding to babies, 11
Penguins, 37

# R

Rattlesnakes, 75
Record Evidence Like a Scientist, 40–42, 94–97
Responding, to sound, 34–36, 37, 47

# S

Senses, exploring sound with, 64–67
Sheep, 21–23
Ships, 44
Sirens, 54–55
Social behavior, 33
Sorting sounds, 58
Sound energy, 64–67, 87
Sound Exploration Center, 64–66
Sounds
   body parts to hear, 82
   describing, 68
   determining how they are made, 56–57
   different ways to make, 59–61
   heard by ears, 86
   helping animals survive, 8
   importance of, 12
   jobs making, 98–101
   made with animal parts, 50
   made by animals, 28–29, 96
   made by babies, 11
   made by dogs, 53
   made by humans, 97
   making, 89, 96
   reasons for making, 31–32, 34
   responding to, 34–37, 47
   sorting, 58
   travel of, 87, 92
Sound waves, 88
Strumming, of guitar strings, 99

# T

Tails, 75
Telephone
  making, 88–93
  sound traveling through, 92
Think Like a Scientist, 19–25, 64–67
Throats, 75
Traits, inheritance of, 14
Trunks, of elephants, 76

# V

Vibrations
  making, 78–81
  from musical instruments, 69–73
  music as, 99
  and sound, 60
Volume, at concerts, 100

# W

Waves, 88
Whistle, 55
Wings, 74
Wolves, 2, 32